THUMBELINA

THUMBELINA

The Culture and Technology of Millennials

Michel Serres

Translated by Daniel W. Smith

ROWMAN & LITTLEFIELD
INTERNATIONAL
London • New York

Published by Rowman & Littlefield International, Ltd.
Unit A, Whitacre Mews, 26-34 Stannery Street, London SE11 4AB
www.rowmaninternational.com

Rowman & Littlefield International, Ltd. is an affiliate of Rowman &
Littlefield
4501 Forbes Boulevard, Suite 200, Lanham, Maryland 20706, USA
With additional offices in Boulder, New York, Toronto (Canada), and
Plymouth (UK)
www.rowman.com

Originally published in French as *Petite Poucette* © Editions Le
Pommier 2012

British Library Cataloguing in Publication Information Available
A catalogue record for this book is available from the British Library

ISBN: HB 978-1-78348-070-8
ISBN: PB 978-1-78348-071-5

Library of Congress Cataloging-in-Publication Data
Serres, Michel.
[Petite poucette. English]
Thumbelina : the culture and technology of millennials / Michel Serres ; translated by
Daniel W. Smith.
pages cm.
ISBN 978-1-78348-070-8 (cloth : alk. paper) – ISBN 978-1-78348-071-5 (pbk. : alk.
paper) – ISBN 978-1-78348-072-2 (ebook)
1. Information society–Social aspects. I. Title.
B2430.S463P5813 2014
303.48'3–dc23
2014029856

♾TM The paper used in this publication meets the minimum requirements of American
National Standard for Information Sciences Permanence of Paper for Printed Library
Materials, ANSI/NISO Z39.48-1992.

Printed in the United States of America

For Hélène,
the teacher of Thumbelina's teachers
who listens to those who listen to her

For Jacques, a poet,
who makes them sing

CONTENTS

ABOUT THE AUTHOR

Michel Serres is a professor in the history of science at Stanford University and a member of the Academie Francaise. A renowned and popular philosopher and one of France's leading intellectuals, he is the author of a number of works already translated into English, including *Variations on the Body*, *The Parasite*, *The Five Senses* and *Malfeasance: Appropriation Through Pollution?*

I

THUMBELINA

Before teaching anything to anyone, we should at least know who our students are. Who, today, is enrolling in our schools, colleges and universities?

NOVELTIES

This young schoolgirl and new schoolboy have never seen a calf, a cow, a pig, or a brood of chicks. In 1900, most human beings on the planet worked the land; by 2011, in France and in similar countries, the number of people working the land had been reduced to one percent of the population. This has been one of the greatest revolutions in history since the Neolithic period. Our cultures, which used to be tied to pastoral practices, have suddenly changed. Of course, what we eat, on the planet, still comes from the earth.

The children I am introducing here no longer live in the company of animals; they no longer inhabit the same earth; they no longer have the same relation to the world. The nature they admire is merely an Arcadian nature, the nature of vacations and tourism.

They live in the city. More than half of their immediate predecessors haunted the fields. Yet, having become aware of the environment, these prudent and respectful students pollute less than we unconscious and narcissistic adults.

They no longer endure the same physical life, nor are they living in the same numerical world. The world's population, during the span of a single human life, has leapt from two to almost seven billion human beings. They are living in a world that is packed full.

Their life expectancy is now close to eighty years. Their great-grandparents, on their wedding day, had promised to be faithful to each other for perhaps a decade. If he and she decide to live together, will they swear to the same fidelity for sixty-five years? Their parents received their inheritance around age thirty, while they will have to wait until old age to receive their legacy. They no longer have the same life-span, nor the same marriage, nor the same transmission of patrimony.

Their parents, heading off to war with a flower in their gun, often sacrificed their short life spans to their motherland. Will these children honor themselves in the same way when they have the promise of six more decades ahead of them?

For the past sixty years, an atypical period in the history of the West, none of these children has ever known war firsthand, nor have their leaders or teachers.

Statistically speaking, they have suffered much less than their predecessors, having reaped the benefits of a medicine that has finally become efficacious, notably through the development of analgesics and anesthetics. Have they known hunger? Every morality, whether religious or secular, was embodied in exercises aimed at dealing with inevitable and daily pain: sickness, famine, the cruelty of the world.

They no longer have the same body or the same behavior; adults no longer have any hope of inspiring in them even an adapted morality.

Their parents were conceived blindly, whereas their birth was programmed. Since the average age of a woman having her first child has increased by ten to fifteen years, the parents of these students come from a different generation. For more than half these children, their parents are divorced from each other. Have they left their children as well?

Neither he nor she have the same genealogy.

Their predecessors met in classes and lecture halls that were culturally homogeneous, whereas they study in a collectivity where they mingle with students from numerous religions, languages, origins, and customs. For them, and their teachers, multiculturalism is the rule. How long will they still be able to sing, in France, of the "impure blood" of foreigners?[1]

They do not inhabit the same global world; they do not inhabit the same human world. All around them, they encounter the sons and daughters of immigrants from less affluent countries, who have lived through vital experiences that are the opposite of theirs.

A provisional assessment: they are fortunate. They know nothing of the rustic life, domestic animals, or the summer harvest. They have not lived through ten wars, the wounded, the starving, the motherland, bloody flags, cemeteries, or monuments to the dead. Nor have they ever experienced, through their suffering, the vital urgency of a morality. What literature or history will they be able to understand?

FROM THE BODY TO KNOWLEDGE

The culture of their ancestors was grounded in a temporal horizon of several thousand years, adorned with Greco-Latin antiquity, the Jewish Bible, a few cuneiform tablets, and a short prehistory. This temporal horizon has now been extended billions of years, going back to the Plank barrier, and passing through the accretion of the planet, the evolution of the species, and a paleontology spanning millions of years.

No longer inhabiting the same time, they are living a completely different history.

They are formatted by the media, which is broadcast by adults who have meticulously destroyed their faculty of attention by reducing the duration of images to seven seconds, and the response time to questions to fifteen seconds—these are official figures. The word that is repeated most often in the media is "death," and the most frequently represented images are those of corpses. In the first twelve years of their lives, these adults will force them to watch more than twenty thousand murders.

They are formatted by advertising. How can French children be taught that the word "relais" is written with an "-ais," when it is spelled with an "-ay" in every train station in France?[2] How can they be taught the metric system when the SNCF, the French national rail system, decides to call its fidelity program "S'Miles"—one of the most ridiculous things in the world?

We adults have transformed our society of the spectacle into a pedagogical society whose overwhelming competition, willfully ignorant, has eclipsed the school and the university. The media long ago took over the function of teaching—the time when one hears and sees, the time of seduction and consequence.

In France, our professors—criticized, mistrusted, and lambasted, since they are poor and discreet, even if they hold the world record for recent Nobel prizes and Field medals in relation to the size of the population—have become far less influential than these dominant teachers of the media, who are boisterous and rich.

These children inhabit the virtual. The cognitive sciences have shown us that using the Internet, reading or writing messages (with one's thumb), or consulting Wikipedia or Facebook does not stimulate the same neurons or the same cortical zones as does the use of a book, a chalkboard, or a notebook. They can manipulate several forms of information at the same time, yet they neither understand it, nor integrate it, nor synthesize it as do we, their ancestors.

They no longer have the same head.

With their cell phone, they have access to all people; with GPS, to all places; with the Internet, to all knowledge. They inhabit a topological space of neighborhoods, whereas we lived in a metric space, coordinated by distances.

They no longer inhabit the same space.

Without us even realizing it, a new kind of human being was born in the brief period of time that separates us from the 1970s.

He or she no longer has the same body or the same life expectancy. They no longer communicate in the same way; they no longer perceive the same world; they no longer live in the same Nature or inhabit the same space.

Born via an epidural and a programmed pregnancy, they no longer fear, with all their palliatives, the same death.

No longer having the same head as their parents, he or she comprehends differently.

He or she writes differently. After watching them, with admiration, send an SMS more quickly than I could ever do with my clumsy fingers, I have named them, with as much tenderness as a grandfather can express, Thumbelina (*Petite Poucette*) and Tom Thumb (*Petit Poucet*). These are their real names, much nicer than the old pseudo-scientific French word, *dactylos* (typists).

They no longer speak the same language. Since the time of Richelieu, the French Academy has published its own dictionary, every twenty years or so, as a reference work. In prior centuries, the difference between the two publications was four to five thousand words, a fairly constant measure.

Between the current dictionary and the next one, the difference will be around thirty-five thousand words.

With this rhythm, it is obvious that our descendants will soon find themselves as distant from our language as we are, today, from the Old French spoken by Chrétien de Troyes or Jean de Joinville. This gradient provides a quasi-photographic sign of the changes I am describing.

This immense difference, which affects most languages, is derived in part from the changes between the jobs of recent years and those of today. Thumbelina and her friend will no longer have to apply themselves to the same kind of work.

Language has changed, and work has mutated.

THE INDIVIDUAL

Better yet, they have all become individuals. The individual was invented by St. Paul at the beginning of our era, but it has been born again in our time. Until recently, we lived in what we might call our "belongings": French, Catholics, Jews, Protestants, Muslims, atheists, Southerners or Northerners, females or males, poor or rich. . . . We belonged to regions, religions, cultures (rural or urban), teams, towns, a sex, a dialect, a party, and a motherland. Through travel, images, the web, and abominable wars, almost all these collectives have broken apart.

Those that remain are unraveling.

The individual no longer knows how to live in a couple; it divorces. It no longer keeps to its own kind; it moves about and chats with anyone and everyone; it no longer prays in its own parish. In the summer of 2011, the French soccer players no longer knew how to be a team. Do our politicians still know how to construct a plausible party or a stable government? Everyone speaks of the death of ideologies, but what is disappearing is rather the *belongings* recruited by these ideologies.

This newly-born individual is good news. When I weigh the harm done by what grumpy old men call "egoism" against the crimes committed by and for the *libido* of belongings—hundreds of millions of deaths—I love these young people to death.

That being said, new links still need to be invented, as evidenced by the number of people who use Facebook, quasi-equivalent to the world's population.

Like an atom without valence, Thumbelina is completely naked. We adults have not invented any new social links; our generalized tendency toward suspicion, critique, and indignation had led instead to their destruction.

Rarely seen in history, these transformations, which I call *hominescent*, have created, in our time and in the midst of our own collectivities, a rupture so large and obvious that few have measured its magnitude.[3] It is comparable to the more visible ruptures of the Neolithic, the beginning of the Christian era, the end of the Middle Ages, and the Renaissance.

Although we are on the other side of this fault line, we are still teaching our young people in institutional frameworks that come from a time they no longer recognize. Buildings, playgrounds, classrooms, lecture halls, campuses, libraries, laboratories, even forms of knowledge—these frameworks, I am saying, date from a time and were adapted to an era when both the world and humans were something they are no longer.

Three questions, for example.

WHAT TO TRANSMIT? TO WHOM TO TRANSMIT IT? HOW TO TRANSMIT IT?

What to transmit? Knowledge!

Until fairly recently, the support of knowledge was the body of the knower, such as a storyteller or a bard. The body of the pedagogue was itself a living library.

Gradually, however, knowledge became objectivized, first in scrolls or on pieces of vellum or parchment, which were the supports of writing; then, during the Renaissance, in books made out of paper, which found their support in the printing press; and finally, today, on the web, which is the support of email messages and information.

The historical evolution of the support-message couple is a good variable by which to assess the function of teaching. Teaching has changed, rather suddenly, at least three times. With writing, the Greeks invented *paideia*; after the printing press, treatises on pedagogy proliferated. Today?

I repeat. *What to transmit? Knowledge? It is already available and objectivized on the web. Transmit it to everyone? Knowledge is already accessible to everyone. How to transmit it? Done!*

With access to people through cell phones, and access to places through GPS, access to knowledge is now open. In a certain manner, it has already been transmitted, always and everywhere.

Objectivized, certainly, but more importantly, *distributed*. Knowledge is no longer concentrated; it is distributed. We used to live in a metric space, as I said, which was linked to centers or concentrations. A school, a classroom, and a lecture hall are concentrations of people, students and professors; a library is a concentration of books; a laboratory is a concentration of instruments. But now, this knowledge—these reference works, these texts, these dictionaries, and even observatories!—are distributed everywhere, and in particular, in your own home. Even better, you can access them from almost any place you happen to find yourself. You can reach your colleagues or students anytime and anywhere, and they can easily respond.

The old space of concentrations, the space where I speak and you listen, has been diluted and expanded. We are living in a distributed space of immediate neighborhoods. I can speak to you from my home or anywhere else, and you can listen to me from anywhere—even your home. What, then, are we doing here?

Above all, we cannot say that students lack the cognitive faculties to assimilate this distributed knowledge, since these faculties have been transformed along with, and because of, their support. With the advent of writing and the printing press, for instance, memory mutated to the point where Montaigne said he preferred *"une tête bien faite plutôt qu'une tête bien pleine,"* "a good head rather than a full head." This head has now mutated yet again.

Pedagogy was invented by the Greeks (*paideia*) when writing was invented and disseminated, and it was transformed anew in the Renaissance when the printing press appeared. Today, pedagogy is again changing completely with the advent of new technologies, whose novelties are only one variable among the many I have cited or could enumerate.

We all sense that we urgently need a decisive change in teaching, a change that will eventually have repercussions on the entire space of our global society and its obsolete institutions. Indeed, it will not only affect teaching, but also work, business, health, law and politics—in short, all our institutions. Yet this change is still far off, no doubt because those who lived through the transition from these final states have not yet retired, and they are instituting reforms using models that have long since been surpassed.

I have taught, for half a century, in almost every latitude of the world, and everywhere, this crack is opening up as wide as it has been in my own country. In each case, I have been subjected to—and suffered from—these reforms, like a plaster cast on wooden legs, a mere patching-up. But plaster damages the tibia, even an artificial one, and the patching-up tears the very tissue it is meant to strengthen.

Yes, for the past several decades I have watched us live through a period comparable to the dawn of *paideia*, when the Greeks learned to write and demonstrate. Or the Renaissance, which experienced the birth of the printing press and

the reign of the book to come. Yet the two periods are in-
comparable. At the same time as these techniques are mutat-
ing, the body itself is metamorphosing, changing both birth
and death, suffering and healing, occupations, space, habi-
tats, being-in-the-world.

ENVOI

Faced with these mutations, we no doubt need to be invent-
ing unimaginable novelties, far outside the obsolete frameworks that still format our behaviors, our medias, and our
projects—all of which are being drowned in the society of
the spectacle. I see our current institutions sparkling with a
light that is similar to the light of distant constellations,
which astronomers have taught us have long been dead.

Why have these innovations not taken place? I hesitate to
accuse philosophers (I consider myself to be one of them),
although their vocation is to anticipate the knowledges and
practices to come, and it seems to me that they have failed in
this task. Preoccupied with day-to-day politics, they have not
perceived the arrival of the contemporary.

If I had tried to sketch a general portrait of adults (I also
consider myself to be one of them), my profile would have
been much less flattering.

I would like to be eighteen years old, the age of Thumbelina and Tom Thumb, since everything has to be redone,
everything still needs to be invented.

I hope that life leaves me enough time to work on this,
side by side with Thumbelina and Tom Thumb, to whom I
have dedicated my life because I have always respectfully
loved them.

NOTES

1. The reference is to the refrain of *La Marseillaise*, the French national anthem: "To arms, citizens, / form your battalions, / let's march, let's march! / Let an impure blood / water our furrows!"

2. In France, *Relay* is a well-known chain of newspaper, magazine, book, and convenience stores, which are located primarily in train stations and airports. The word *Relay* is an Anglicized variation of the French word *relais*, which originally referred to a way station for changing horses.

3. Serres is referring to his earlier book *Hominescence* (Paris: Le Pommier, 2001), which developed in more detail many of the analyses presented in *Thumbelina*.

2

SCHOOL

THUMBELINA'S HEAD

In his book *The Golden Legend*, Jacobus de Varagine tells the story of a miracle that took place in Lutetia (Paris) during the century of persecutions decreed by the Emperor Decius. The Roman army arrested Denis, who had been elected bishop by the first Christians of Paris. Imprisoned, then tortured, in the Île de la Cité, he was condemned to be beheaded at the top of a hill that would later be named Montmartre.

The soldiers, somewhat lazily, decided not to go to the top, and instead executed the victim half way up the hill. The bishop's head rolled to the ground. Horror! Decapitated, Denis got up, grabbed his head, nestled it in his arms, and continued to climb the slope. Miracle! Terrified, the legionnaire ran off. The author adds that Denis took a break to wash his head, and kept walking until he reached the present location of St. Denis. He was later canonized.

Thumbelina opens her computer. Even if she does not know this legend, she is nonetheless beholding her own head, in front of her and in her hands. It is a full head, because of its enormous stock of information, but it is also a well-made head, since its search engines bring up texts and images at a moment's notice, and its programs process huge amounts of data faster than she could ever do herself. She is holding, outside of herself, a cognition that used to be inside her, just as St. Denis held his head severed from his neck. Has Thumbelina been decapitated? Miracle?

Not long ago, we all became like St. Denis. Our intelligent head has been externalized outside our skeletal and neuronal head. In our hands, the computer-box contains and manages what we used to call our "faculties": a memory thousands of times more powerful than our own; an imagination stocked with millions of icons; and a faculty of reason as well, since software programs can solve hundreds of problems that we could never solve on our own. Our head has been projected before us in an objectified cognitive box.

After the beheading, what is left on our shoulders? An innovative and enduring intuition. The learning process, which has fallen into the box, has left us the incandescent joy of invention. Has this condemned us to become intelligent?

Once the printing press appeared, Montaigne, as I've said, preferred a well-made head to an accumulated knowledge, since this accumulation, already objectivized, could be found in the books on the shelves of his library. Before Gutenberg, you had to know Thucydides and Tacitus by heart if you practiced history; Aristotle and the Greek mechanics, if you were interested in physics; Demosthenes and Quintilian, if you wished to excel in public speaking. You needed to have a full head. Economize: remembering the location of a volume on the shelf of the library costs less in memory than retaining its entire contents. Now we have a new, more radical, economy: we no longer even need to remember the location, since a search engine does all the work.

Thumbelina's severed head, better made than filled, is very different from her mother's. Since she no longer has to work hard to gain knowledge—it is already in front of her, objective, collected, collective, connected, accessible at her leisure, already reviewed and edited—she can return to the absence that hovers over the severed neck. There, she will find the air, the wind, and—even better—the light portrayed by Léon Bonnat, the academic painter, when he painted the miracle of St. Denis on the walls of the Pantheon in Paris. There, she will find the new genius, the inventive intelligence, an authentic cognitive subjectivity. It is as if her originality takes refuge in this translucent emptiness, in this cool breeze. Knowledge at almost no cost, yet difficult to grasp.

Is Thumbelina presiding over the end of the era of knowledge?

THE HARD AND THE SOFT

How did this decisive change in humanity come about? Practical and concrete, most of us think, almost inevitably, that revolutions occur around hard things. What is most important to us are tools, such as hammers and sickles. We have even named various periods of history after them: the recent Industrial Revolution, the Bronze Age and the Iron Age, the ages of cut stone and polished stone. Almost blind and deaf, we pay far less attention to signs (the soft) than to tangible machines (the hard and practical).

Yet far more than tools, it was the invention of writing—and, later, printing—that truly revolutionized our cultures and collectivities. The hard shows its effectiveness in the things of the world; the soft shows its effectiveness in the institutions of humans. Techniques lead to or assume the hard sciences; technologies presuppose and lead to the humanities, public meetings, politics and society.[1] Without writing, would we have gathered together in cities, stipulated laws, founded states, conceived of monotheism and history, invented the sciences, and established *paideia*? Would we have ensured their continuity? Without printing, would we have modified, during the aptly named Renaissance, all these institutions and assemblies? The soft organizes and federates those who utilize the hard.

Though we hardly realize it, we are living together, today, as the children of the book and the grandchildren of writing.

THE SPACE OF THE PAGE

Because of the printing press, writing is now projected everywhere in space, to the point where it has invaded and even hidden the landscape. Advertising billboards, traffic signs, street signs, schedules at train stations, scoreboards in stadiums, translations at the Opera, scrolls of the prophets in synagogues, the Gospels in churches, blackboards in classrooms, PowerPoint in lecture halls, magazines and newspapers. . . . The *page* dominates us and guides us. And the screen reproduces the page.

Consider maps of cities, rural land registries, architect's blueprints, plans for construction projects, or designs for public halls and private bedrooms. Each of these mimic, in their soft and paginated grid-patterns, the *pagus* of our ancestors: square parcels of earth planted with alfalfa, or plots of furrowed land, on whose hardness the peasant left the trace of his plow. A furrow is already a line being written within this demarcated space. Such is the spatial unity of perception, action, thought, and intentionality. This is a format that has persisted for millennia, almost as important to humans, or at least to Westerners, as the hexagon is to bees.

NEW TECHNOLOGIES

This page-format so dominates us—though we are hardly aware of it—that even our new technologies have not been able to break away from it. The screen of a laptop computer—which itself opens like a book—mimics the page, and Thumbelina still writes on the screen with ten fingers, or with two thumbs on her smart phone. Once she is finished writing, she can immediately print out her work. Innovators in every domain are heralding the new electronic book, but the electronic has not yet been liberated from the book, even though it implies something completely different from the book and the trans-historical format of the page. What that thing is has yet to be discovered. Thumbelina can help us here.

I remember my astonishment when, a few years ago, on the campus of Stanford University, where I have taught for thirty years, a building was erected in steel and concrete, with glass windows, near the old Quadrangle, devoted to information technology and financed by billionaires from nearby Silicon Valley. In effect, it was identical to the older brick buildings on campus where mechanical engineering and medieval history had been taught for more than a century: the same layout of the floors, the same classrooms, the same hallways. Always the same format inspired by the page. As if the recent revolution, at least as powerful as those of the printing press and writing, had changed nothing in our knowledge, or in our pedagogy, or in the space of the university itself, which was invented long ago by and for the book.

No. The new technologies are forcing us to leave the spatial format implied by the book and the page.

A SHORT HISTORY

First: everyday tools externalize our forces and capacities; they are hard. Our muscles, bones, and joints leave the body, as it were, and are fitted out as simple machines—levers and hoists—which mimic their functioning. Our high temperature, the source of our energy, comes out of our organism, and is fitted out as motor machines. The new technologies, finally, externalize the messages and operations that circulate in our neuronal system—information and codes, which are soft. Cognition, in part, is fitted out in this new tool.

What then remains, today, above the cut necks of St. Denis's sons and daughters?

THUMBELINA MEDITATES

Cogito: my thinking is distinguished from knowledge, and from the various processes of understanding—memory, imagination, deductive reason, discernment, geometry—that have been externalized, along with synapses and neurons, in the computer. Moreover, I can think and invent better if I keep this knowledge or understanding at a distance from myself, if I open up a gap between it and myself. I *become* this emptiness, this impalpable air, this *soul*—a word that translates this "wind." I think in a way that is softer than this objectivized softness. I can invent if I manage to reach this emptiness. I can no longer be recognized by my head, or by the density of what fills it, or by its singular cognitive profile, but rather by its immaterial absence, by the transparent luminosity that lingers after its decapitation. By this nothingness.

If Montaigne had tried to elucidate the ways a head could be marvelously made, he would have been thinking of a compartment to be filled and would have returned to the idea of a full head. If we were to describe this empty head today, it would fall outside of us in the computer. No, we cannot cut off one head simply to replace it with another. We cannot experience anxiety when faced with this emptiness. Keep going, be strong. Knowledge and its formats, understanding and its methods—not to mention the infinite details and admirable syntheses that my predecessors amassed like protective armor in footnotes at the bottom of their pages, and in massive bibliographies at the end of their books, and which they have accused me of omitting—all this, in the wake of the sword cut of St. Denis' torturers, has fallen into the electronic box. Strangely, quasi-instinctively, the *ego* recoils from this fall, from this leap into the void, into its white and innocent nullity.

The subject of thinking has changed. The neurons activated in the white fire of the cut neck differ from those that writing and reading activated in the heads of their predecessors, and which now drive the computer.

Hence the new autonomy of our understanding, which finds its complement in corporeal movements without constraint and a brouhaha of voices.

VOICES

Until recently, teachers, in their classrooms or lecture halls, would communicate to their students a knowledge that could already be found in books, at least in part. They oralized writing, a page-source. If they invented—a rare thing—they could write up their thoughts afterwards, a page-compilation. It was their academic standing that made us listen to their voice, and they demanded silence whenever they delivered their oral lectures. They no longer get it.

Starting in childhood, in elementary and secondary schools, the wave of what we call "chatting," which becomes a tsunami in colleges, has now reached even graduate schools, whose lecture halls, overflowing with chat, are filled, for the first time in history, with a permanent brouhaha, which makes it difficult to listen. One can hardly hear the old voice of the book. This is a rather general phenomenon, and we must pay attention to it. Thumbelina does not read, nor does she want to have repeated to her what has already been written—like the old advertisement about a dog who no longer listens to the voice of its master. Having been reduced to silence for almost three millennia, Thumbelina, along with her brothers and sisters, is now producing, like a choir, a background noise that drowns out the voice of writing.

Why does she chat, in the midst of the hubbub of her chatting comrades? Because she already has the knowledge that was promised to her by the university. All of it. Whenever she wants it. In her hands. It is accessible, from any portal, via the web, from Wikipedia. It is explained, documented, and illustrated, with no more errors than those found in the best encyclopedias. The voice of yesteryear is no longer needed—unless someone, original and rare, invents.

It is the end of the era of knowledge.

SUPPLY AND DEMAND

This new chaos, primitive like every disorder or brouhaha, heralds a complete change—first, in pedagogy; and then, in every aspect of politics. Until recently, teaching consisted in supplying a demand. It was an exclusive offer, dispensed by those who had little interest in listening to the advice or choices of those making the demand. Knowledge is stocked in the pages of books: thus spoke the Teacher-Voice, who read the books, then delivered and demonstrated the knowledge. Listen to me first, then you can read, if you want. In any case, keep quiet.

Twice, the supplier said: Be quiet!

All that is over. As its wave has grown, chatting has refused this supply in order to invent, proclaim, and implement a new demand, which no doubt will require another knowledge. This is the reversal! Those of us who are Speaking Teachers must learn how to listen, in our turn, to the confused and chaotic noise of this chattering demand coming from our students. Until recently, no one bothered to ask if they were truly demanding what we were supplying.

Why is Thumbelina less and less interested in what the Teacher-Voice has to say? Because, given the increasing supply of knowledge from in an immense depository—everywhere and always accessible—a one-time and singular supply becomes ridiculous. The problem was posed rather cruelly when, to obtain this rare and secret knowledge, you had to travel long distances. Now accessible, knowledge is overabundant; and it is always within reach, in the small volume that Thumbelina carries around in her pocket. The wave of access to knowledge rises as high as the wave of chatting.

Supply without demand is now dead. The enormous supply that has succeeded and replaced it is immediately available to those making the demand. This is already true in our schools, and I suggest that it will soon become true in politics. Is this the end of the era of experts?

ENTRANCED CHILDREN

With its eyes and ears fixed on the Teacher-Voice, the dog, seated, is so fascinated by what it is hearing that it never moves. At a tender age, always on our best behavior, we likewise began a long career of sitting motionless on our bottoms, in silence and in rows. Our new name: *Petits Transies*, Entranced Children. With our empty pockets, we obeyed. We were subject not only to our masters, but even more so, to knowledge, which our masters themselves humbly obeyed. Both they and we assumed that knowledge was sovereign and magisterial. No one would have thought to write a treatise on our voluntary subjection to it. Some were even terrorized by knowledge, and were thereby prevented from learning. They were not stupid, but terrified. We must try to grasp this paradox: knowledge wanted to be received and understood, but for that reason it needed to terrify.

Philosophy sometimes even spoke of Absolute Knowledge, in capital letters. It asked that we bow our heads in reverence to it, much as our ancestors were cowed by the absolute power of kings with divine right. A democracy of knowledge has never existed, not because those who had knowledge possessed power, but because knowledge itself required humiliated bodies, including the bodies of those who possessed knowledge. That most self-effacing of bodies, the teaching body, gave courses that genuflected to this absent absolute, this inaccessible totality. Entranced, the other bodies did not move.

Already formatted by the page, the spaces of our schools, colleges, and campuses were reformatted by this hierarchy, which was directly inscribed in the body's posture. Silence and prostration. The bodies were all focused on the podium, where the Teacher-Voice demanded silence and immobility. Sit down, sit up, and sit still. This pedagogical hierarchy is reproduced in the focus of the courtroom on the judge, the theatre on the stage, the royal court on the throne, the church on the altar, the habitation on the hearth . . . of the multiplicity on the one. Narrow seats, aligned in rows, for the immobilized bodies of these cave-institutions. Such was the tribunal that condemned St. Denis. Is this the end of the era of actors?

THE LIBERATION OF BODIES

The innovation: the ease of access that has been given to Thumbelina, and to the entire world, their pockets filled with all the knowledge in their smart phones. Their bodies can finally leave the cave, where attention, silence, and their curved spines once bound them to their chairs like chains. Even if we force them to go back, we can no longer sit still in their seats. They have become, as they say, disruptive.

No. The space of the lecture hall was designed as a field of forces whose orchestral center of gravity was the stage, with its focal point at the lectern, which was literally a *power point*. What was situated there was the heavy density of knowledge, which scarcely existed on the periphery. Now, knowledge is distributed everywhere, moving freely in a homogeneous and decentered space. The classrooms of yesteryear are dead, even if we still see classrooms everywhere, even if they are the only thing we know how to build, even if the society of the spectacle is still trying to inflict them upon us.

But bodies have now been set in motion. They circulate, they gesticulate, they call and question each other, and they willingly exchange what they have found in their smart phones. Has chatting replaced silence, or have disruptions replaced immobility? No. Once a prisoner, Thumbelina has simply freed herself from the thousand-year-old chains that shackled her to her seat, immobile and silent, mouth closed, firmly in her place.

MOBILITY: DRIVER AND PASSENGERS

The centered or focused space of the classroom or lecture hall can be likened to the inside of a vehicle—a car, a train, an airplane—where the passengers, seated in rows, allow themselves to be driven by the person piloting them to knowledge. But take a look at the bodies of the passengers: they are slouched in their seats, their stomachs protruding, with vague and passive looks. By contrast, the driver is active and attentive, his back straight, his hands grasping the steering wheel.

When Thumbelina uses her computer or smart phone, these devices both require the body of a driver, alert and active, and not that of a passenger, relaxed and passive. Demand and not supply. Her back is straight, her stomach is taut. Thrust such a person back into a classroom, and her body, used to driving, will no longer tolerate the posture of a passive passenger. Deprived of a machine to drive, she becomes active and busy in other ways. A disturbance. Put a computer in her hands, and she will reassume the gestures of the body-pilot.

Today, nothing remains but drivers, nothing but motricity. With no spectators, the space of the theater is filled with actors, who remain mobile. In the courtroom, there are no judges, only speakers and orators, who are constantly active. In the sanctuary, there are no priests; in church, everyone is a preacher. In the lecture hall, there are no masters; everyone has now become a professor. And, we must add, the political arena is no longer occupied by the powerful, but by

the people on whose behalf the powerful used to make deci-
sions.

It is the end of the era of the decision-maker.

THE INSTRUCTED THIRD

Thumbelina seeks and finds knowledge in a machine. Rarely accessed, this knowledge used to be presented in bits and pieces, cut up and divided. Page after page, scholarly classifications distributed to each discipline its lot: its departments, its localities, its laboratories, its slice of the library, its credibility, its Teacher-Voices and their corporate structure. Knowledge was divided into sects, and reality disintegrated into pieces.

A river, for example, was dispersed in the scattered basins of geography, geology, geophysics, hydrodynamics, the crystallography of alluvia, the biology of fish, halieutics, and climatology, not to mention the agronomy of irrigated plains, the history of flooded cities, the rivalry between riverside residents, and then the bridge, the gondolier's barcarole, and even Apollinaire's *Le Pont Mirabeau*. Our easy access to knowledge—by bringing together this debris, merging and integrating it, restoring from these scattered limbs the living body of the current—allows us, finally, to inhabit the river in full.

But how can we unite these classifications, dissolve these borders, gather together the already cut and formatted pages, superimpose the designs of the university, unify the lecture halls, pack up the departments in one suitcase, and make all the high-level experts—each of whom thinks they possess the exclusive definition of intelligence—listen to each other. How can we transform the space of the campus, which mimics the space of the defensive camp of the Roman army, both of which are traversed by the usual paths, and divided up into juxtaposed cohorts or lawns.

Answers: by listening to the background noise coming from the demand, from the world and its populations; by following the new movements of their bodies; by trying to explicate the future implicated in the new technologies. But once again: How?

THE DISPARATE VERSUS THE CLASSIFIED

Put differently: How, paradoxically, can we produce Brownian movements? We can at least promote them through the serendipity of Aristide Boucicaut.

The founder of the French department store *Le Bon Marché*, Boucicaut initially tried to classify all the merchandise he wanted to sell, arranging it neatly on shelves in separate aisles. Each package sat quietly in its place, classified, organized, like students in rows or Roman legionaries in their entrenched camp. Indeed, the term "class" originally signified the division of an army into tight rows. Now, since Boucicaut's huge store—which was as universal for the *Bonheur des dames*[2] as the university was for the pleasure of learning—had assembled under one roof everything a regular customer could dream of (food, clothing, cosmetics), success came quickly, and Boucicaut made a fortune. Zola's novel about this inventor recounts Boucicaut's rude awakening when his revenues, one day, hit a ceiling and refused to budge.

One morning, struck by a brash intuition, he destroyed his reasonable ordering and instead turned the aisles of his store into a labyrinth and its shelves into chaos. You might come to the store to buy leeks for a soup, but because of this vigorously programmed chaos, you first had to pass through the section selling silks and laces. So Thumbelina's grandmother wound up buying a tablecloth as well as her vegetables. The sales went through the ceiling.

The ill-sorted or the disparate has virtues of its own, of which reason is unaware. Order, though practical and efficient, can imprison. Although it promotes movement, in the end it can also freeze movement. The *check-list*, though essential for action, can sterilize discovery. An atmosphere penetrated with disorder, by contrast, is like an apparatus that has a certain play in it, and it is precisely this play that provokes invention. And this same play appears between the neck and its severed head.

Let us follow Thumbelina as she plays her computer games. Let us listen to Boucicaut's serendipitous intuition, which has been put into practice by every subsequent department store. Let us undo the orderings of the sciences, and put the Department of Physics next to Philosophy, Linguistics alongside Mathematics, and Chemistry with Ecology. We even need to work on the details, altering, for instance, the contents of our menus. Every time users start their computers, they could encounter new menus, as if from a strange planet, speaking another language. They would thus travel far without ever leaving home. The rational *castrum* of the Roman army, marked out by perpendicular lines and separated into square groups, would be replaced by a mosaic of diverse bits and pieces, a kind of kaleidoscope, the art of marquetry, a potpourri.

The Troubador of Knowledge[3] was already dreaming of universities whose space was mixed and multi-colored, striped like a tiger, blended in different shades, dyed with numerous pigments, twinkling like the stars—real like a landscape! We used to have to travel to see other people, or we could stay at home to avoid them; now they are in our face, non-stop, without our having to move.

People whose work defies classification, or who sow seeds into every wind, perpetually fertilize inventiveness. Pseudo-rational methods, by contrast, are good for nothing. How can we redesign the page? By forgetting the order of reasons. We need order, certainly, but an order without reason. It is reason that must be changed. The only authentic intellectual act is invention. Our preference should be for the labyrinth of electronic chips.

Long live Boucicaut and my grandmother! cries Thumbelina.

THE ABSTRACT CONCEPT

And what should we think about concepts, sometimes so
difficult to form? Tell me, for instance, what Beauty is.
Thumbelina answers: a beautiful woman, a beautiful dance,
a beautiful sunset. . . . Stop! Why, when I ask you about a
concept, do you instead give me a million examples? When
will you stop playing with your dolls and horses?

Every abstract idea brings with it an immense economy of
thought. Beauty holds in its hands a thousand and one beautiful women, just as the geometer's circle includes an infinite
myriad of round things. We could never have written or read
neither pages nor books if we had to cite each of these beau-
ties or circularities, since their number is enormous, without
limit. Moreover, I would never be able to demarcate even a
single page without appealing to an idea that would halt this
indefinite enumeration. Abstraction functions like a cork
stopper.

Do we still need it? Our machines now scroll so quickly
that they are able to count particularities indefinitely—they
know how to prevent originality. If the image of light can
again be used to illustrate—dare I say—knowledge, our ancestors chose it for its clarity, whereas we opt rather for its
speed. Abstraction can be replaced, at least sometimes, by a
search engine.

Just as the subject has changed, so has the object of cognition. We do not have an ineluctable need for concepts.
Sometimes we need them, but not always. We can linger as
long as necessary in narratives, examples, and singularities—

the things themselves. Practically and theoretically, this innovation restores dignity to the knowledge of description as well as the individual. As a result, knowledge bestows its dignity upon the modalities of the possible and the contingent—to singularities. Once again, a certain hierarchy collapses. Having become experts in chaos, mathematicians can no longer distrust geology and biology, which were already sciences of *mélange* or mixture in the manner of Boucicaut, and thus had to teach in an integrated fashion. If a living reality is cut up in an analytic manner, it dies. Once again, the order of reasons—still useful, certainly, but often obsolete—gives way to a new reason that welcomes the concrete singular, naturally labyrinthine, into the narrative.

The architect destroys the divisions of the campus.

The space of circulation, a diffuse orality, free movements, the end of classified classes, disparate distributions, the serendipity of invention, the speed of light, the novelty of both subjects and objects, the search for a new reason. . . . The diffusion of knowledge can no longer take place on any campus in the world, which are themselves ordered and formatted by the page, rational in the old manner, imitating the camps of the Roman army. This is the space of thought where Thumbelina, in both her body and her soul, spent her youth.

St. Denis pacified the legion.

NOTES

1. The French language distinguishes between *techniques* and *technologies*. Very generally, a technique is a process or practice of fabrication, whereas technology (from the Greek *tekhnè*, technique, and *logos*, discourse or study) is a discourse about techniques. In English, the word technology has expanded to cover almost all senses of the Greek *tekhnè*.

2. Emile Zola's 1883 novel, *Au Bonheur des Dames* ("For the Happiness of Women"), which was set in a store modeled after *Le Bon Marché*. It has been translated into English as *The Ladies' Paradise*, trans. Brian Nelson (Oxford: Oxford University Press, 1995).

3. A reference to Michel Serres' book *The Troubador of Knowledge*, trans. Sheila Faria Glaser and William Paulson (Ann Arbor: University of Michigan Press, 1997), whose original French title was *Le tiers-instruit* (Paris: François Bourin, 1991).

3

SOCIETY

IN PRAISE OF RECIPROCAL GRADES

Should Thumbelina grade her teachers? A curious debate on this question raged in France not long ago. From afar, I was astonished. For the past forty years, students at other universities have been grading me. I didn't come off badly. Why? Because, even without being required to do so, anyone who takes a course is always evaluating their professor. On the first day of class, a crowd of students crammed into my classroom; today, only three or four are left: I have been penalized by numbers. Or by attention: students either listen or become bored, distracted, and disruptive. Cause of itself, eloquence finds its origin in the listener's silence, itself born of eloquence.

In fact, everyone everywhere is always getting graded: a lover, by his silent mistress; a merchant, by the shouts of his customers; the media, by ratings; a doctor, by the influx of patients; elected officials, by the sanction of voters. This brings us, quite simply, to the question of government.

The obsession with grading, encouraged by anxious mothers and psychologists, quickly left the schools and invaded civil society, which at every opportunity publishes lists of bestsellers, hands out Nobel Prizes, Oscars, and various Cups in false metal, ranks universities, rates banks, appraises companies and even states—and before them, kings and queens. In turning this page, dear reader, you will at that very moment be evaluating me as an author.

A kind of two-faced demon compels us to judge something to be good or bad, innocent or harmful. True lucidity, however, discerns what is dying in the old world and emerging in the new. Today, a reversal is occurring that promotes a symmetrical circulation between the graders and the graded, a reciprocity between the powerful and the subjugated. Most people seem to believe that everything flows from top to bottom, from the pulpit to the pews, from the elected to the electors. Upstream, the supply is presented; downstream, the demand swallows everything. Large megastores, great libraries, powerful CEOs, ministers, and statesmen, presuming our incompetence, generously bestow their beneficent rain on we little people. Perhaps such a situation existed at some point, but it has ended in our lifetimes—at work, in the hospital, on the road, in groups, in public places, everywhere.

Freed from these supposed drivers—by which I mean these asymmetrical relations—the new circulation is making us listen to the sound, quasi-musical, of its voice.

IN PRAISE OF HUMPHREY POTTER

It has been said that Humphrey Potter, a boy from Birming-
ham, used the string of his top to tie together, on the arm of
a steam engine, the inlet and exhaust valves that he had been
hired to activate by hand. Trying to avoid a boring job so he
could play, Potter invented, while suppressing his slavery, a
kind of feedback loop. True or not, this tale is meant to
praise the precociousness of a genius. To my eyes, it shows,
rather, the frequent competence, subtle and adapted, of
workers, even minor ones, in places where the deciders, distant, have issued commands to act without consulting their
actors, who are assumed to be incompetent. Humphrey Potter is one of the *noms de guerre* of Thumbelina.

The word *employee* expresses this presumption of incompetence. Etymologically (*employer*, from the French *plier*, to
fold), it is a matter of folding something at one's leisure in
order to exploit it. A disease is reduced to an organ to repair;
a student, to an ear to fill or a silent mouth to force-feed.
Workers are reduced to machines to be managed, only
slightly more complicated than the machines they work on.
Above, there used to be deaf mouths; below, mute ears.

Let us sing the praises of reciprocal control. By restoring
complete faces on both levels, the best companies put their
workers at the center of practical decision making. Far from
organizing, like a pyramid, a logistics of flows and a regulation of complexity (which in fact multiplies complexity by
adding layers of regulation), they let Thumbelina control her
own activity in real time. Breakdowns are more easily iden-

tified and repaired, technical solutions are found more quickly, productivity is improved. But we must also examine her representatives—here, her bosses; and in a moment, doctors and politicians.

THE DEATH OF WORK

Thumbelina is searching for a job. Even when she finds one, however, she will still be searching, since she knows that, tomorrow, she could easily lose the job she has just found. Moreover, at work, she answers those who speak to her, not in terms of the question posed, but in a way that will not make her lose her job. Now a commonplace, this lie harms everyone.

Thumbelina is bored at her job. The carpenter next door used to obtain untreated boards from a mill in the forest and, after letting them dry, would turn this treasure into stools, tables, or doors, depending on his customers' orders. Thirty years later, he gets ready-made windows from a factory, which he installs in pre-fabricated walls with formatted openings. He's bored. So is she. The stimulating work takes place in the office of the designers, far from the construction site. Capital does not simply signify a concentration of money, but also water behind a dam, ore under the earth, and the intelligence of a group of engineers who work far from those who execute their plans. The boredom comes from this concentration, this inveiglement, this theft of interest.

Jobs have become increasingly scarce, due to increased productivity, which has grown dramatically since 1970, as well as the growth of the global population, which has added to productivity. Will an aristocracy soon be sole beneficiary of this productivity? Born in the Industrial Revolution, and copied from the divine offices of the monasteries, is work dying today, slowly but surely? Thumbelina has watched the number of blue collar workers diminish; the new technologies will cause the number of white collar workers diminish as well. Will work disappear as much as its products—products that inundate the markets and often harm the environment, contaminated by the action of machines and the fabrication and transportation of goods? Work depends on sources of energy whose exploitation depletes their reserves and pollutes.

Thumbelina dreams of a new kind of work that would right these wrongs and be a benefit to those who work. She is not simply speaking of her salary—otherwise she would have said "beneficiary"—but of her happiness as well. She draws up a summary list of actions that would help prevent two kinds of pollution: not only the pollution of the planet, but also the pollution of human beings. The French utopian thinkers of the nineteenth century, distrusted for being dreamers, attempted to organize human practices in directions diametrically opposed to those that have led us to this double impasse.

Since there are only individuals, and since society has organized itself around work, and since everything is made to revolve around work—even encounters and private adventures that have nothing to do with it—Thumbelina had hoped to find fulfillment in her job. But in fact she hardly finds anything at all in it. She is bored. She tries to imagine a society that would no longer be structured by work. But by what?

And how often has she been asked for her advice?

IN PRAISE OF THE HOSPITAL

She also remembers her stay at a large hospital. The doctor entered her room without knocking, followed, like a dominant male, by submissive females (the model of animals is hard to avoid). He gratified his herd with a high-flying discourse, while turning his back on Thumbelina, who lay in bed and received the presumption of incompetence. Just like at school, or at work. Put crudely, we could say she was treated like an imbecile.

The imbecile, from the Latin *imbecillus*, is a lame person, someone who lacks a staff (*bacillus*) to hold himself up. Upright and healed, Thumbelina proclaims the good news, like a new version of Oedipus's riddle: the more time advances, the less humanity needs this staff as its third leg. It can stand on its own two feet.

Listen. Public hospitals in our great cities have parking areas for wheelchairs and gurneys: for emergencies; before and after an MRI or other forms of scanning; before an operation, for anesthesia; afterward, for the recovery. Patients can wait in these areas from one to six hours. To scientists, to the rich and powerful of the world, I say: Do not avoid these places. You will hear suffering, pity, anger, despair, tears, sometimes prayer, exasperation, the supplications of those who call out to those who do not call out and who deplore the absence of a response, the tense silence of some, the alarm of others, the resignation of most, and gratitude as well. Those who have never had to add their voice to this dissonant concert no doubt know they are suffering, but they will never know what "we are suffering" means, the shared lamentation that emanates from this antechamber of death, an intermediate purgatory that everyone dreads, and where one can only hope for a reprieve from fate. If you have ever asked yourself the question, "What is humanity?" it is here, in the midst of this quiet tumult, that you will receive, learn, and perhaps even give the response. Until he hears this, even a philosopher is literally absent-minded.

Such is the noise of the depths, the human voice that murmurs beneath our discourses and chatter.

IN PRAISE OF HUMAN VOICES

This chaos does not simply reverberate in our schools and hospitals. It does not simply emanate from Thumbelinas in class, or through tears in a waiting room. Teachers talk among themselves when a principal announces a decision; interns converse when a physician holds forth; policemen speak when a sergeant commands; citizens, assembled in a marketplace, start heckling when a mayor, deputy, or minister starts to stonewall. Show me, says Thumbelina, ironically, a single gathering of adults that does not produce, amusingly, a similar brouhaha.

These real voices are covered over by muzak, the din of the media, the deafening and anaesthetizing racket of commerce—shameful noise and calculated drugs—not to mention the virtualities of blogs and social networks, whose sum total, almost incalculable, approaches numbers comparable to the population of the planet. Yet for the first time in history, the voice of almost everyone can be heard. Human speech vibrates in space and throughout time. Small and silent villages, whose calm was broken, rarely, by bells and clocks—law and religion, the sons and daughters of writing—were replaced by the vast and noisy expanse of these networks. This is such a widespread phenomenon that we need to pay attention to it. It is the new noise of the depths, a cacophony of clamoring voices—private, public, permanent, real or virtual—a chaos that includes all the outdated motors and electronic devices of the society of the spectacle. It reproduces on a large scale the mini-tsunami of chatter that has invaded our classrooms and lecture halls. Or rather, the latter is a reduced model of the former.

Do these thumb-chats, these cacophonies of the world, herald an era in which a second oral age will join hands with virtual writings? Will the waves of this novelty wind up drowning the age of the page that has formatted us? For a long time now, I have heard this new oral age emanating from the virtual.

This general demand to be able to speak is analogous to the singular demand that Thumbelina articulates in schools and universities. It is the expectation of hospital patients and employees at work. Everyone wants to speak, everyone now communicates with each other through countless networks. This tissue of voices harmonizes with the network of the web; they are in sync with each other. This new democracy of knowledge already exists in places where the old pedagogy has exhausted itself and the new one is being sought, with as much persistence as there is difficulty. In general politics, it corresponds to a democracy-in-formation that, soon, will become inescapable. The political supply, concentrated in the media, will die; while the political demand, which is enormous, will rise up and exert its pressure, even if it cannot yet express itself, and indeed does not know how. The voice used to cast its vote on a paper ballot, narrow and perforated, local and secret. Today, its ever-present noise occupies the totality of space. The voice is permanently casting its vote.

IN PRAISE OF NETWORKS

On this very point, Thumbelina reproaches her fathers. You accuse me of being egotistic, but who showed me how? Or for being individualistic, but who taught me how? Look at you—do you know how to be part of a team? You can't even live together as couples; you're constantly divorcing. Do you know how to create and sustain a political party? Everywhere, parties are weak and diminished. Can you constitute a government that lasts? Can you play collective sports? To enjoy the game, you recruit players from distant countries where people still know how to live and act as a group.

All the old *belongings* are dying: brotherhoods of weapons, parishes, motherlands, unions, and even families, which are being reorganized. All that remains are the lobbyists, shameful obstacles to democracy.

You make fun of our social networks and our new use of the word "friend." Have you ever managed to form groups so large that their number approaches that of humanity itself? Is there not a certain prudence in first approaching others virtually to avoid hurting their feelings? No doubt, you are afraid that these endeavors will lead to new political forms that will sweep away the preceding forms, now obsolete.

Obsolete, indeed, and as virtual as my own, retorts Thumbelina, suddenly animated. Army, nation, church, people, class, proletariat, family, market . . . these are abstractions, flying above our heads like cardboard fetishes. Incarnated, you say? Certainly, she responds, except that this human flesh, instead of living, has to suffer and die. Bloodthirsty, these belongings expected each of us to offer their life as a sacrifice. Look at the Church and its laws: tortured martyrs, stoned women, heretics burned alive, accused witches immolated on pyres. Or the State: unknown soldiers laid to rest by the thousands in military cemeteries, where an occasional dignitary will study, with solemnity, the long lists of names on the monuments of the dead. In 1914–1918, almost the entire peasantry of France was killed. Look at the mad theory of "races" and the class struggle, which produced the extermination camps and gulags. As for the family: it shelters half of all criminal acts. Every day, a woman dies from the abuse of a husband or lover. And as for the market: a third of humanity suffers from hunger—a Thumbelina dies from it every minute—while the elites are dieting. Even the amount of aid given to others, in your society of the spectacle, grows with the number of cadavers that are reported, and the sales of your novels, with the crimes that are narrated. For you, good news is not news. For the past hundred years, we have been tallying these deaths, of all kinds, in the hundreds of millions.

Rather than these belongings, whose names are abstract virtualities and whose bloody glory is praised by the history books; and rather than these false gods who consume infinite victims, I prefer our immanent virtual which, like Europe, does not require anyone's death. We no longer want to coagulate our assemblies with blood. The virtual, at least, avoids this charnel house. We will no longer build collectivities on the massacres of others. This is *our* future, after what we have seen of your history and your politics of death.

Thus spoke Thumbelina, still alive.

IN PRAISE OF TRAIN STATIONS AND AIRPORTS

Listen, she said, to the rustle of the soft crowds passing by. *Homo sapiens* used to be constantly on the move, depending on available game and fruits, and variations in climate. For a long time, *Homo sapiens* was *Homo viator* (traveling man), until the day came, relatively recently, when the planet no longer contained any unknown lands. Since the development of various kinds of motors, traveling has become so common that the perception of our habitat has been transformed. A country like France quickly became a city, with the high-speed trains (TGVs) functioning like a subway, and its super-highways like its streets. By 2006, the world's airlines had transported a third of humanity. So many people pass through airports and train stations that they are starting to resemble transient motels.

Considering the number of times she has moved, does Thumbelina know what city she lives and works in, or to which community she belongs? She lives in a suburb of the capital, though sometimes it takes more time to travel to the city center or the airport than it does to reach the nearest border. She thus resides in a metropolitan area that extends beyond her city and her nation. Question: Where does she live? Both reduced and expanded, this *place* raises a political question, since the word "politics" refers to the city. But of what city is she a citizen? Another fluctuating belonging! Who, coming from where, will represent her—she who puts in question the very place of her habitation?

Where? At school; in the hospital, in the company of people from many countries; at work, or commuting with strangers; at a meeting with translators; walking down the street, where she hears several languages. She rubs shoulders with a human melting pot, which perfectly reproduces the blend of cultures and knowledges she encountered during her education. For the reversals we are describing have also affected the population density of the world's countries, where the West is shrinking compared to the rising tide of Africa and Asia. Human mixtures flow like rivers to which we ascribe proper names, but their waters combine dozens of tributaries. Thumbelina lives in a composite tapestry and covers her space with a disparate marquetry. Her eyes are fascinated by this kaleidoscope, and her ears ring with a confused chaos of voices and sensations that foreshadow other reversals.

REVERSAL OF THE PRESUMPTION
OF INCOMPETENCE

Using the old presumption of incompetence, the great public and private machines—bureaucracies, the media, advertising, technocracy, businesses, politics, universities, government, and sometimes even the sciences—mete out their immense power by speaking to supposed imbeciles, who are
dubbed the "general public," distrusted by these various networks. But the anonymous Thumbelinas of the world have
joined forces with others whom they presume to be competent, even if they are not so sure of themselves. Their diverse
voices are a sign that these dinosaurs, which become noisier
the more they are headed for extinction, are completely unaware of the emergence of these new competencies. Here's
why.

If she consults a good website, Thumbelina (which in fact
is a code name for a student, a patient, a worker, an employee, an administrator, a traveler, a voter, a senior or an adoles-
cent or whatever, a child, a consumer, in short, that anonymous person of the public arena that we sometimes call a
"citizen") can know as much, or more, on a given subject—a
decision to be made, public information, the care of one-
self—as any teacher, manager, journalist, senior official,
CEO, or president, all of whom have reached the pinnacle of
the spectacle and are obsessed with glory. Oncologists often
say they learn more from reading blogs written by women
with breast cancer than from their many years in medical
school. Specialists in natural history can no longer ignore

what people say online, whether they are Australian farmers talking about the habits of scorpions or guides in the Pyrenees discussing the movement of chamois. This kind of sharing reintroduces symmetry into education, health care, and work. Listening accompanies discourse, and the inversion of the old iceberg promotes circulation, in all senses of the word. The *collective*, whose virtual character was hidden, fearful, under a monumental death, gives way to the *connective*, which is truly virtual.

At the end of my studies, when I was twenty years old, I became an "epistemologist," which is a big word to say that I studied the methods and results of the sciences, and occasionally tried to judge them. There were, at that time, very few of us, so we corresponded with each other. A half-century later, every Tom Thumb on the street can make judgments about nuclear power, surrogate mothers, GMOs, chemistry, and ecology. Though I myself no longer claim to work in the discipline, today everyone has become an epistemologist. There is a *presumption of competence*. Don't laugh, says Thumbelina. When democracy gave everyone the right to vote, it did so in opposition to those who considered it a scandal to give an equal vote to both wise men and fools, ignorant and educated. The same argument applies here.

The great institutions I have just mentioned still constitute the entire décor of what we continue to call our "society," even though society has been reduced to a stage that, every day, is losing more of its plausibility. It hardly even attempts to rejuvenate the spectacle, crushing an increasingly perceptive populace with an ever greater mediocrity. These great institutions, I repeat, resemble those stars whose light we receive, but which astrophysics calculates have been dead a long time. For the first time in history, no doubt, the "public"—those individuals and persons we used to call "commoners," in short, Thumbelinas—can and will be able to possess as much wisdom, science, information, and capacity for decision-making as these institutional dinosaurs, whose voraciousness for energy and avarice in production we still serve as subjected slaves. Like mayonnaise, these solitary monads will organize, slowly, one by one, to form a new body, with no relation to these solemn and lost institutions. When this slow constitution suddenly turns over, like the aforementioned iceberg, we will pretend we had no clue it was about to happen.

This reversal affects the sexes, since recent decades have witnessed the victory of women, who are more assiduous and serious at school, in the hospital, and in business, which are nonetheless dominated by males, arrogant and weak. That is why this book is entitled *Thumbelina*. It also affects cultures, since the web promotes a multitude of expressions and, soon, automatic translation, even though we have barely emerged from an era where the massive domination of a single language had unified speech and thought in mediocrity, sterilizing innovation. In short, it affects all concentrations—whether productive and industrial, linguistic, or even cultural—to the benefit of large, multiple, and singular distributions.

Here, finally, is a generalized form of grading. Here is a generalized vote for a generalized democracy. The conditions are ripe for a Western spring, although the opposing powers-that-be have now resorted to drugs instead of force. An example from everyday life: things have lost their common names and instead are known by the proper names of brands. This has happened with every form of information, including politics, which is staged in well-lit arenas where politicians fight shadows unrelated to any reality. The society of the spectacle transforms the struggle—which was once (and is elsewhere) hard, with barricades and corpses—into a heroic detoxification that would cleanse us of the sedatives distributed to us by the many providers of lethargy . . .

IN PRAISE OF MARQUETRY

. . . who, to preserve the old state of things, have recourse to an argument that appeals to simplicity: How can we manage the complexity of the voice and its brouhaha, which is disparate and composite, a disorder? Here's how. Caught by a net, a fish tries to swim away, but more it wriggles to get free, the more it get ensnared; flies, by vibrating their wings, become imprisoned in the spider's web; mountaineers, meeting on a rock face and confronted with danger, mix up their ropes even more if they try to disentangle them too quickly. Administrators sometimes issue guidelines to reduce administrative complexity, but they thereby multiply it, imitating the alpinists. Is complexity a state of things in which any attempt to simplify it complicates it further?

How can we analyze this? Through the growth of the number of elements, their individual differentiation, the multiplication of their relations, and the intersections of their trajectories. Graph theory and computer science deal with these figures of crossed networks, which topology calls a *simplex*. In the history of the sciences, the appearance of complexity was a sign that you had utilized an inadequate method and needed to change your paradigm.

It is this type of connected multiplicity, however, that characterizes our societies, which lie at the intersection of individualism, the demands of persons or groups, and the mobility of sites. Today, everyone weaves their own simplexes, and move about in others'. Just now, Thumbelina was moving about in a space that was mixed and striped . . . in a labyrinth, before a mosaic of kaleidoscopic colors. Freedom belongs to every individual and demands that they have free hands and elbow room, and no one thinks this requirement of democracy should be simplified. Indeed, simple societies lead to animal hierarchies, subject to the law of the strongest: a pyramidal body with a single summit and a wide base.

May complexity proliferate, finally! Yet it comes at a cost: waiting lines multiply and become longer; administrative red tape increases; there is congestion in the streets, and difficulties in interpreting sophisticated laws, whose density in fact diminishes freedom. We always pay in the currency we earn.

Moreover, this cost appears to be one of the sources of power. It is one reason citizens suspect that their representatives have no desire to reduce these complications. Their numerous directives may give the impression they want to reduce it, but instead the complexity is multiplied, like fish caught in a net.

IN PRAISE OF THE THIRD SUPPORT

The history of the sciences, I repeat, has familiarized us with the transformations that accompany this type of growth. When the old Ptolemaic model had accumulated so many epicycles that the movement of the stars became complicated and unreadable, a change became necessary. The center of the system was moved toward the sun, and everything became clear again. The written code of Hammurabi no doubt put an end to the socio-juridical difficulties that had arisen in oral law. Our own complexities come from a crisis of writing. Laws multiply, the *Journal official* expands. The page has run its course. Things must change, and it is the computer that provides the relay. We used to wait and get jostled in lines at store counters. Caught in endless traffic jams, you might even run over your father at an intersection, unknowingly, in an altercation over the right-of-way. Electronic speed, however, avoids the delays of real transportation, and the transparency of the virtual annuls collisions at intersections, along with the violence they entail.

May complexity never disappear! It is growing, and will continue to grow, because everyone benefits from the ease and freedom it provides. It is the primary characteristic of democracy. To reduce its cost, it is enough to *will* it. Engineers can solve this problem using computers, whose capacities conserve the simplex and even allow it to grow, but process it quickly, thereby, I repeat, suppressing queues and traffic jams, and eliminating collisions. The development of an appropriate software for a virtual passport, incorporating personal and publishable data, would take a few months, no more. Someday, all these data will have to be placed on a single new platform. At the moment, they are distributed in various cards, whose ownership the individual shares with various public and private institutions. How long will Thumbelina—an individual, a customer, a citizen—allow states, banks, and megastores to appropriate her personal data, which have today become such a source of wealth? This a political, moral, and legal problem, and it is transforming our historical and cultural horizon.

The result may be a realignment of our socio-political landscape by the advent of a fifth power—the power of data—which is independent of the four other powers: legislative, executive, and judicial powers, and the media.

What name will Thumbelina have printed on her passport?

IN PRAISE OF THE *NOM DE GUERRE*

The name of my heroine does not indicate "someone of her generation" or "a teenager of today," which are expressions of contempt. No. There is no question here of selecting an element *x* from a set *A*, as one says in set theory. Unique, Thumbelina exists as an individual, as a person, not as an abstraction. This is worth explaining.

Who even remembers the old division, in France and elsewhere, between the four faculties: letters, sciences, law, and medicine? The first of these sang the glories of the *ego*, the personal "I," the human being of Montaigne, and of historians, linguists and sociologists. The faculties of the sciences, in describing, explaining, and calculating the *it*, expressed general, even universal laws—Newton, for the equation of celestial bodies, Lavoisier, in the naming of bodies. But medicine and law, perhaps without realizing it, have both attained a mode of knowledge that neither letters nor science were capable of accessing. Uniting the general and the particular, juridical and medical faculties gave birth to a third subject, who was one of the ancestors of Thumbelina.

First, her body. Until fairly recently, what one saw on an anatomy chart was a schema—a schema of the hip, the aorta, the urethra—an abstract, quasi-geometrical, and general drawing. Today, one can examine an MRI of the hip of *this* eighty-year-old man, or the aorta of *that* sixteen-year-old girl. Even though they are individual, these images have a generic and qualitative significance. Roman jurisconsults and casuists, in studying a case, were in the habit of naming it after a subject cited in the case, such as Gaius or Cassius. These code names, along with pen names, *noms de guerre*, and pseudonyms, function simultaneously on two levels, forming a bridge between the general and the particular. They are doubles, so to speak, that apply to both.

We should understand Thumbelina to be a code name for this student, this patient, this worker, this peasant, this elector, this passerby, this citizen . . . *anonymous*, certainly, but *individuated*. She is not a quantity—like a voter who counts for one in opinion polls, or a TV viewer who counts for one in the Nielsen ratings—but rather a quality, an existence. Like the unknown soldier of old—whose body truly lies here and could be individuated by a DNA analysis—this anonymous person is the hero of our time.

Thumbelina codes this anonymity.

ALGORITHMIC, PROCEDURAL

Now, let us watch Thumbelina manipulating her cell phone
and mastering the thumb-buttons on the search engines or
games. Without hesitation, she is deploying a cognitive field
that part of the previous culture, the culture of sciences and
letters, had long left fallow, which could be called "procedural." In elementary school, such sequences of hand movements were primarily used to learn, correctly, the simple
operations of arithmetic, and also, perhaps, as rhetorical or
grammatical artifices. Today, such procedures have penetrated all our knowledge and techniques. They are poised to
compete with abstract geometry, as well as the descriptions
of the sciences that are without mathematics. They form an
algorithmic mode of thought. Through it, we have begun to
comprehend the order of things and to improve our practices. It used to be a part of juridical practice and medical
art, although somewhat blindly. Neither of these were taught
in our colleges of letters and sciences, precisely because they
made use of recipes, rules of order, sequences of gestures,
series of formalities, yes, *procedures*.

Consider: the landing of aircrafts on busy runways; the
connections between air, rail, automotive, and maritime
routes on a given continent; a long surgical operation on a
kidney or heart; the merger of two industrial companies; the
solution of an abstract problem whose demonstration re-
quires hundreds of pages; the use of GPS. These all involve
an activity that is very different from geometrical deduction
or experimental induction. The objective, the collective, the

technological, the organizational: today, these depend far more on this *algorithmic or procedural cognition* than the *declarative* abstractions to which philosophy, nourished by letters and sciences, has dedicated itself for more than two millennia. Merely analytic, philosophy today has not been able to see this mode of thought coming into being. It cannot *think* it, since it lacks not only the means, but also its objects, and indeed, its subject. It has missed our time.

EMERGENCE

Yet this novelty is not new. Algorithmic thought arose before the invention of geometry in Greece, and reemerged in Europe with Pascal and Leibniz, who invented two calculating machines and, like Thumbelina, used pseudonyms. Formidable, but at the time discreet, this revolution was not noticed by the philosophers, still nourished by sciences and letters. Between geometric formality of the sciences and the lived reality of letters, this revolution brought about a new cognition of things and humans which had already been foreseen in the practice of medicine and law, both of which united the universal and the particular, jurisdiction and jurisprudence, the sickness and the sick. Out of this, our own novelty emerged.

Since then, thousands of efficacious methods have utilized algorithms or procedures. The culture of Leibniz and Pascal—a direct inheritance of the Fertile Crescent before Greece, and of al-Khwārizmī, a Persian scientist writing in Arabic—has today invaded the domain of the abstract as well as the concrete. Letters and sciences have lost an ancient battle that began, as I have said elsewhere, with the *Meno*, one of Plato's dialogues, where Socrates the geometer misunderstands a young slave who, instead of making a demonstration, utilizes procedures. I have named his anonymous servant Tom Thumb: he has won out over Socrates. After two millennia, this marks the return of the presumption of competence.

The new victory of these old procedures stems from the fact that the algorithmic and the procedural both rely on codes. So let us return to names.

IN PRAISE OF CODE

"Code" is a term (*codex*) that has always been used in law and jurisprudence, as well as in medicine and pharmaceutics. Today, biochemistry, information theory, and the new technologies have taken up the word, and thereby generalized it to knowledge and action in general. In the past, most people knew little about juridical codes or the codes of medication. Both open and closed, the writings of law and medicine, though published and public, could be read only by the learned. A code resembled a coin with two sides, heads and tails, which were contradictory: both accessible and secret. Only recently have we begun to live in a civilization of access. The corresponding linguistic and cognitive element of our culture is the code—that which permits or prohibits access. Since the code establishes a set of correspondences between two systems that must be translated into each other, it possesses the two faces we need for the free circulation of flows, whose novelty I have been describing. Coding preserves anonymity, while permitting free access.

Now, the code is a singular, living being; the code is a concrete human being. As an individual who is unique but also generic, who am I? *An indefinite, decipherable, and undecipherable cipher*, open and closed, social and discreet, accessible-inaccessible, public and private, intimate and secretive. I am sometimes unknown to myself and on display at one and the same time. I exist, therefore I am a code. I am calculable and incalculable, like a golden needle, plus the haystack in which, buried, its brightness lies hidden. My

DNA, for example, is both open and closed; its cipher has constructed my body, which is both intimate and public, like St. Augustine's *Confessions*. How many signs are there in the *Confessions*? How many pixels in the *Mona Lisa*? How many bytes in Faure's *Requiem*?

Medicine and law have long been sustained by this idea of the human being as a code. Today, this has been confirmed by both knowledge and practice, whose methods make use of *procedures* and *algorithms*. The code gives rise to a new *ego*. Personal, intimate, secret? Yes. Generic, public, publishable? Yes. Better, both at once: double, like the pseudonym.

IN PRAISE OF THE PASSPORT

Like us, the ancient Egyptians distinguished the body of a human being from its soul, but they added to this duality a double, named Ka. Certainly, we can reproduce the body, from the outside, by science, screens and formulae. And we can describe the intimate soul through *Confessions*, like those of Rousseau (how many signs?). Can I likewise reproduce my double, which is accessible and publishable, even though it is indefinite and secret? Is it enough to code it? By generalizing all possible data—intimate, personal, and social—the "Carte Vitale," the French health insurance card, for instance, created a Ka, a universally coded passport: open and closed, a double that is public and secret, without contradiction. Is there anything more strange? Although I try to think on my own, I speak in a common language.

This *ego* can quietly confess to itself, in its soul and consciousness, but it can also be slipped into one's pocket, in a hard plastic case. Subject, yes; object, yes; thus, once again, a double. Double like a patient, singularly in pain, but offered up, like a landscape, to the medical gaze. Double, competent and incompetent . . . double like a citizen, public and private.

THE IMAGE OF SOCIETY TODAY

In an unforgettable time in the past, certain heroes wanted to work together to build a high tower. However, since they came from disparate lands and spoke untranslatable idioms, they failed to reach their goal. No comprehension, no possible team. No collective, no building. The Tower of Babel barely gets off the ground. Thousands of years passed.

As soon as prophets or scribes started to write, in Israel, Babylon, or around Alexandria, so many teams became possible that the pyramid started to rise, as well as the temple and the ziggurat. They were built. Thousands of years passed.

One beautiful morning in Paris, a human gathering called the *Exposition universelle*—World's Fair—gave rise to a similar project. An expert head drew up a blueprint on his paper and, after choosing the materials, calculated their resistance and riveted together lattices of steel until they reached more than 300 meters in height. Since then, the Eiffel Tower has kept watch over the left bank of the Seine.

From the Egyptian pyramids to the Eiffel Tower, the first in stone, the second in iron, the global form has remained stable. In a stable state, and stable like the State: the two words are equivalent. Static equilibrium is inseparable from the model of power, which remains invariant through a thousand apparent variations, whether religious, military, economic, financial, or specialist. It is a power always held by those, on high, who are closely bound by money, an armed force, or another apparatus capable of dominating a base that is large and low. From the monster of rock to the dinosaur of iron, there has been no notable change. The form may be more open, transparent, and elegant in Paris, more compact and squat in the desert, but in both cases, it remains pointed at the top and wide at the base.

The democratic decision changes nothing in this schema. Sit in a circle, on the ground, and you will all be equals, said the ancient Greeks. Cunningly, this lie pretends not to see, at the base of the pyramid or the Tower, the *center* of the assembly, which marks the projection of the summit of the pyramid onto the ground, the place where its sublime pinnacle touches land. This is the democratic version of centralism, the Communist party used to say, which merely repeats this old illusion of stagecraft, even though in the neighboring center sat Stalin and his henchmen, who deported, tortured, and killed. Lacking real change, we, subjects of the periphery, prefer a distant power, at the top of the axis, to this terrifying neighbor.

Our French forebears launched their revolution less against the king, who was still rather popular, than to eliminate the evil baron next door.

Chéops, Eiffel, the same State.

Michel Authier, the brilliant designer, along with myself, his assistant, plan to light a fire or plant a tree across from the Eiffel Tower on the right bank of the Seine. On computers, located on-site and elsewhere, people will scan their passport—their Ka, an anonymous and individuated image, their coded identity—and a laser beam, brilliant and colored, will emerge from the ground and reproduce the innumerable sum of these documents, thereby projecting a virtual image of the collectivity, created virtually. Every participant, of their own accord, will become a part of this virtual and authentic team, which will unite, in a unique and multiple image, the individuals that belong to this disseminated collectivity, with their concrete and coded qualities. Within this high icon, as high as the Tower, the common characteristics will be assembled in a kind of trunk, the rarest in branches, and the exception in foliage or buds. But since the sum is constantly changing, and will be transformed from day to day, from one second to another, the tree thus raised will vibrate wildly, as if set ablaze by dancing flames.

Thus, in front of the immobile iron Tower that proudly bears the name of its creator—and forgetful of the thousands that built it, some of whom died in the process—and in front of the Tower that supports, at its summit, a transmitter broadcasting the voice of the master, a new tower will start its dance, a tower that will be variable, mobile, fluctuating, colorful, striped, clouded, inlaid, mosaic, musical, and kaleidoscopic, a tower that will speak in flying sparks of chromatic light, representing the connected collective, all the more real for being virtual and participative, constructed from the data of each person—who will themselves decide what they want the tower to be. Volatile, living, and soft, today's society releases a thousand tongues of fire from the monster of old— hard, pyramidal, and ice-cold. Dead.

Babel, the oral stage, no tower. From the Pyramids to Eiffel, the written stage. A tree in flames, a perennial novelty.

I am enchanted with you, Thumbelina, but I must be severe. If you remain in Paris, you will quickly become old, both of you. We must fire up this volatile tree on the banks of the Rhine as well, so that its image can also dance for my German friends; and high on the Col Agnel, so it can sing with my Italian colleagues; and along the beautiful blue Danube, on the banks of the Baltic. . . . Truths on this side of the Mediterranean, the Atlantic, and the Pyrenees, and truths beyond, toward the Turks, the Iberians, the North Africans, the Congolese, the Brazilians . . .

TRANSLATOR'S ACKNOWLEDGMENTS

I owe a debt of gratitude to Ashley Albrecht, who read a draft of the translation and offered numerous suggestions for improvement, as well as Justin Litaker, who corrected the final manuscript and suggested the subtitle. I would also like to express my thanks to Sarah Campbell at Rowman & Littlefield, who has been a constant source of support and insight.

ABOUT THE TRANSLATOR

Daniel W. Smith is professor of philosophy at Purdue University. He is the author of *Essays on Deleuze* and translator of several works, including Deleuze's *Francis Bacon*.